MYSTERIES OF THE ANDROMEDA GALAXY

By James Sterling

Andromeda's spiral

Text copyright © 2021 James Sterling

All Rights Reserved

Table of Contents

1. INTRODUCTION

2. THE DISCOVERY

3. CREATION OR EXPLOSION

4. THE MYSTERIES

5. AT THE SPEED OF LIGHT

6. TO INFINITY AND BEYOND

7. EPILOGUE

Chapter 1
Introduction

Allow me take you on an adventure to the stars. These stars, or more eloquently put, celestial bodies, have allowed us to expand our minds and allow our imaginations to grow; captivating us for untold centuries. Science fiction authors have written about our sister galaxy, Andromeda, and movie studios bring her to life on screen through television and cinema.

In this book we will explore the various mysteries of this vast collection of stars, planets, and other nebulous clusters. The goal of this text is to inspire you to accomplish, contemplate, create, and contribute to the advancement of the human race.

Space is vast...

Chapter 2
The Discovery

First off, let us discuss a bit of history and explore how this galaxy was first discovered.

In the year 964 C.E., a Persian astronomer known as Abd al-Rahman al-Sufi first discovered what he called a "nebulous smear" in a book he wrote at that time. This later came to be known as the Andromeda galaxy, a sister to our own galaxy, the Milky Way.

In later years, in the year 1612 C.E. in fact, a German astronomer known as Simon Marius first observed the Andromeda galaxy through an early version of a telescope. This device was much weaker than the telescopes we have today, which makes his finding all the more impressive.

The Andromeda galaxy was first drawn in 1850 C.E. by an Anglo-Irish astronomer whose name was William Parsons. Of special note was the spiral nature of the cosmic dust and matter which makes up this galaxy.

Earth

Chapter 3
Creation or Explosion

Now let us go back even further, into the ancient past, way back roughly ten billion years ago, when the Andromeda galaxy was believed to have been formed. No one knows exactly what cosmic forces were strong enough to create the Andromeda galaxy, or our own Milky Way galaxy for that matter.

Some believe in the Big Bang, a massive explosion that created our universe, throwing matter and energy hurling outward with cosmic force from an epicenter, a focal point of untold magnitude. In this theory, the galaxies and stellar bodies we see through our lenses and eyes are constantly in a state of moving further and further apart at an increasing rate.

Our planet Earth and all of the universe surrounding her is constantly moving outward from that first explosion – even if we are unable to perceive it.

Still others follow their religious beliefs, that a higher being created us and all we see around us. Both viewpoints have merit, since we humans do not know everything about the universe in which we live. Our curious nature of asking how and why makes us sentient but not omniscient or all-knowing.

Our potential as a species is infinite.

Chapter 4
The Mysteries

On a stranger note - as we continue our exploration of the mysteries of our sister galaxy - we first received a mysterious radio transmission from Andromeda, captured in 1950 C.E. by a British astronomer named Hanbury Brown. Cyril Hazard, another fellow astronomer, assisted with this enigma and both astronomers worked at the Jodrell Bank Observatory, located at the University of Manchester in Lower Withington, United Kingdom.

Efforts to capture signals from other galaxies and stellar bodies continues to this day. SETI tracks and monitors our skies, holding out hope that we are not alone in the universe.

On August 15th, 1977 scientists picked up a strange series of radio signals emanating from a source beyond our planet. The scientist was so excited to see the strength and pattern of these signals that he wrote Wow! on a sheet of paper displaying these metrics. This became known as the Wow signal and astronomers have been listening for a repeat of this signal ever since that day.

This radio signal was confirmed to not be of Earth origin; the significance of which is that it could very possibly have come from another galaxy.

Andromeda's spiral

Chapter 5
At the Speed of Light

Science and space exploration have exploded in the last hundred years at an unprecedented rate never seen before on planet Earth. We have a long way to go, however, as we currently do not possess the technology or capability to travel to another galaxy.

Some possible theories on future space travel involve several technologies that have not been physically created as of yet. One form would be giant nuclear- powered engines, capable of sustaining power production for long periods of time. Technically our ability to harness nuclear power exists, but nowhere near on the scale we would need for intergalactic travel.

A breakthrough such as one involving nuclear fission is one method we would need to generate nuclear power in the required amounts. Another form would be hydrazine- powered engines on an enormous scale, one we have already partially employed in some of our spacecraft which we have previously launched.

The pioneers Voyager and Voyager II, two of our first spacecraft launched from Earth, are still powered to this day by a similar method. The Voyager spacecraft both employed a hydrazine and radio-isotope thermoelectric generator to power their systems. These craft have been operating for over 50 years, both having been launched in the year 1977, in the midst of the U.S. Space Race. These vessels have been so successful that they have effectively left our solar system in recent years, the first to accomplish such a feat.

There are many other theories concerning power generation and known forms of transportation that are out of our reach. As an example, two far-flung modes are using solar powered sails or by using sails that utilize the solar winds thrown out by our star, Sol, or more commonly known as the Sun.

Even further in our future are ways to move through space including hyperspace travel, FTL (further than light) travel, and locating wormholes to pass through. Allow me to break that down for you.

Hyperspace travel is mainly shown today in science fiction novels, tv shows, and movies. This involves using a fictional technology vaguely called a hyperspace generator. This theoretical

form of travel is probably the furthest out of our reach, on par with travel by wormhole or space anomaly.

Faster than light travel, or FTL, is exactly that - traveling faster than the speed of light. This technology also does not currently exist within our means. Moving an object such as a space ship at FTL speed would require enormous amounts of energy to power an engine designed to move at that speed. Theoretically, such a form of power could exist but we have not yet discovered it.

Event horizon of a Wormhole

Traveling through wormholes, also known as an Einstein- Rosen bridge, is a type of space travel that is theorized to exist that would allow for very fast or instantaneous travel between two points in space.

Chapter 6

To Infinity and Beyond

Now that we relatively know how we may eventually travel to Andromeda, the question becomes – is it worth it? The answer, like many others, depends on the person you ask and which situation we are in at the time. Many are concerned with the fact that all of humanity is located on one planet; leaving us open to a calamity such as an asteroid from space. If the human race was multi- planetary that lowers our risk of being wiped out in one shot a considerable amount. That alone, in many astronomers' and scientists' opinion, is worth the cost and the risk to venture out among the stars.

The cost of such an endeavor would likely require a large number of the nations of our planet to come together in a historic space race type situation such as the one we experienced in the 1960s and 1970s.

The reward, the payoff, in a scientific sense alone would be worth any cost we could imagine. The likelihood, once we can travel at such speeds, that we could find a life bearing planet in our sister galaxy is astronomically greater than the chance of us not finding one. One thing we are noticing as we continue to make discoveries is that the universe loves variety. The current belief, through observation, is that many planets and moons are rocky and metallic, but that some contain water or the ability to have water on their surface. As far as we are aware, water seems to be the limiting factor for discovering life.

Galactic Core

Chapter 7

Epilogue

I hope this small taste of what awaits us out amongst the stars inspires you and fills you with excitement. The young minds reading this text will be the ones who decide if we stay on Earth or truly become a space- faring civilization.

Therefore, I would personally like to thank each and every one of you in advance for your future contributions in letting our species become one we can be proud of and look back on in a positive way for all time. It really is a privilege to exist, to have the power to create such greatness if you only have the will – and time - to do so.

-James

Images used on the pages listed below are courtesy of NASA and JPL pursuant to the public domain free usage licenses and NASA/JPL commercial image licenses. No promotion or endorsement from NASA or JPL is intended.

Link to NASA policy:

https://www.nasa.gov/multimedia/guidelines/index.html

Link to JPL policy:

https://www.jpl.nasa.gov/jpl-image-use-policy

Image pages: Cover, 2, 6, 8, 11, 14, 23

All other images and text are copyright

© 2021 James Sterling

www.ingramcontent.com/pod-product-compliance
Lightning Source LLC
Chambersburg PA
CBHW051833210526
45473CB00005B/1859